SpringerBriefs in Applied Sciences and Technology

SpringerBriefs present concise summaries of cutting-edge research and practical applications across a wide spectrum of fields. Featuring compact volumes of 50 to 125 pages, the series covers a range of content from professional to academic.

Typical publications can be:

- A timely report of state-of-the art methods
- An introduction to or a manual for the application of mathematical or computer techniques
- A bridge between new research results, as published in journal articles
- A snapshot of a hot or emerging topic
- An in-depth case study
- A presentation of core concepts that students must understand in order to make independent contributions

SpringerBriefs are characterized by fast, global electronic dissemination, standard publishing contracts, standardized manuscript preparation and formatting guidelines, and expedited production schedules.

On the one hand, **SpringerBriefs in Applied Sciences and Technology** are devoted to the publication of fundamentals and applications within the different classical engineering disciplines as well as in interdisciplinary fields that recently emerged between these areas. On the other hand, as the boundary separating fundamental research and applied technology is more and more dissolving, this series is particularly open to trans-disciplinary topics between fundamental science and engineering.

Indexed by EI-Compendex, SCOPUS and Springerlink.

More information about this series at http://www.springer.com/series/8884

James Lee · Keane Wheeler ·
Daniel A. James

Wearable Sensors in Sport

A Practical Guide to Usage
and Implementation

 Springer

James Lee
SABEL Labs, Exercise and Sport Science
College of Health and Human Sciences
Charles Darwin University
Darwin, NT, Australia

Keane Wheeler
Oodgeroo Unit, Chancellery
Queensland University of Technology
Brisbane, QLD, Australia

Daniel A. James
SABEL Labs, Exercise and Sport Science
College of Health and Human Sciences
Charles Darwin University
Darwin, NT, Australia

ISSN 2191-530X   ISSN 2191-5318   (electronic)
SpringerBriefs in Applied Sciences and Technology
ISBN 978-981-13-3776-5   ISBN 978-981-13-3777-2   (eBook)
https://doi.org/10.1007/978-981-13-3777-2

Library of Congress Control Number: 2018965887

This Springer imprint is published by the registered company Springer Nature Singapore Pte Ltd.
The registered company address is: 152 Beach Road, #21-01/04 Gateway East, Singapore 189721, Singapore

Preface

This book is intended to bridge a gap between technical research and the widespread adoption of inertial sensors in biomechanical assessment and ambulatory studies of locomotion. Its purpose is to provide a "no-nonsense" guide to using inertial sensors for those from the sports science disciplines who may be unfamiliar with the terms, concepts and approaches that lead to successful usage. Similarly, for those from a technical disciple such as engineering, it introduces the methodologies from sports science that can provide a window into the usage of sensors in a practical environment that extends well beyond bench-testing.

This book draws upon a combined total of over 40 years experience in the development and usage of wearable sensors in sports science and scientific application to sport.

Darwin, Australia   James Lee
Brisbane, Australia   Keane Wheeler
Darwin, Australia   Daniel A. James

Contents

1 Introduction to Wearable Sensors 1
1.1 Approaches and Usage 1
1.2 GPS and Other Sensors 3
1.3 Trends and Availability 4
References 6

2 Theory and Application 7
2.1 Digitisation 7
2.1.1 Sample Rate 8
2.1.2 Resolution 8
2.1.3 Resolution and Range 10
2.1.4 Calibration 11
References 12

3 Acceleration Components 13
3.1 Inertial Component 14
3.2 Gravity Component 14
3.3 Centripedal 16
References 17

4 Case Studies 19
4.1 Approaches 19
4.1.1 Study Design 19
4.1.2 Recruitment 21
4.1.3 Technology Pilot 21
4.1.4 Pilot 22
4.1.5 Full Study 22
4.1.6 Analysis and Interpretation 23
4.2 Gait (Walking/Running) 25
4.3 Swimming 29
References 37

5 Take-Home Messages... 39
   5.1   Importance of Meaningful Data.......................... 39
   5.2   Importance of Accurate Interpretation 39
   5.3   Future Applications.................................... 40
   5.4   Conclusions... 41
   References... 41

Chapter 1
Introduction to Wearable Sensors

The use of inertial sensors in sporting and other applications where human loco-motion is concerned has grown steadily ever since a spring loaded weight was attached to the body segments to determine its movement characteristics Wong et al. (1981). The rise in popularity can be attributed to several technological trends as well as the increasing sophistication of the field of biomechanics, physiology and performance analysis (James and Petrone 2016). Body-worn sensors or wearable technology is attractive because of the potential to measure human movement unobtrusively, in the ambulatory environment and comparative cost when compared to laboratory-based equipment. The barriers to adoption include that of any new technology into a traditional disciple, the limited penetration due to availability, perceived validation of the technology as well as the skill sets required to use the technologies effectively (Lee et al. 2012).

This monograph has been developed to address these barriers by providing a practical introduction to the technologies, a practical guide to its usage together with case studies to aid in its deployment in applications of interest to the reader. This is presented against a background of its development and likely trends going forward. Like any technology, this work will likely show signs of age eventually, but should prove a helpful companion for the wearable sensor whether they be from an engineering discipline or the sports sciences. Onward!!!

1.1 Approaches and Usage

Athletic and clinical testing for performance analysis and enhancement has traditionally been performed in the laboratory where the required instrumentation is available and environmental conditions can be easily controlled. It is between these two environments that the interplay of environmental validity versus laboratory reliability must be carefully considered as new technologies, such as wearables, emerge. In this environment, dynamic characteristics of athletes are assessed using treadmills, rowing and cycling ergometers and even flumes for swimmers. In general, these machines allow for the monitoring of athletes using instrumentation that cannot be used in the training environment but instead requires the athlete to remain quasi-static, thus enabling a constant field of view for optical devices and relatively constant proximity for tethered electronic sensors, breath gas analysis, etc. Today however, by taking advantage of the advancements in microelectronics and other micro-technologies, it is possible to build instrumentation that is small enough to be unobtrusive for a number of sporting and clinical applications; over time, their acceptance has steadily improved and hence their uptake.

Inertial sensors, commonly comprising of accelerometers and angular rate gyroscopes, measure changes in the linear acceleration and angular acceleration, respectively. More recently the inclusion of magnetometers into wearable devices, while not strictly inertial, have been added as they provide a fixed reference frame. These sensors today are widely applied to the kinematics of a body where they can be used as biomechanical markers of the body activity or to derive linear or angular velocities and thus displacement and angular movements. By knowing the inertial properties of the moving body, forces and moments acting on the body centre of mass and distal segments can be estimated, provided that the inertial properties are well estimated in the case of anthropometric segments (James 2006).

Accelerometers measure acceleration at the sensor itself and typically in one or more axis and are millimetres or smaller in size. In general, a suspended mass is created in the design and has at least one degree of freedom. The suspended inertial mass is thus susceptible to displacement in at least one plane of movement. These displacements arise from changes in inertia and thus any acceleration in this direction. Construction of these devices varies but typically uses a suspended silicon mass on the end of a silicon arm that has been acid etched away from the main body of silicon. The force on the silicon arm can be measured with piezoresistive elements embedded in the arm. In recent years, multiple accelerometers have been packaged together orthogonally to offer multi-axis accelerometry.

Accelerometers measure the time derivative of velocity, and velocity is the time derivative of position. Thus, accelerometers can measure the dynamics of motion and potentially position as well. It is well understood though that the determination of position from acceleration alone is a difficult and complex task (James et al. 2004). Instead, accelerometers are often used for short-term navigation and the detection of fine movement signatures and features (such as limb movement). Accelerometers can be used to determine orientation with respect to the earth's gravity as components of gravity are aligned orthogonal to the accelerometer axis. In the dynamic sports environment, complex physical parameters are measured and observed in relation to running and stride characteristics (Herren et al. 1999), and in the determination of gait (Williamson and Andrews 2001).

Typically, human movement assessment often looks at readily understood quantities— velocity and displacement—to measure performance. These are both relatively easy to measure with a range of tools from simple approaches such as stopwatches and 2D video systems. Early application of sensors attempted with the same mindset of needing to measure these two kinematic outcomes. However, inertial sensors always came up for short of accurate measures. This was largely due to drift in componentry and unwanted noise. When integration is applied to such data, outcomes of exponential error occur. It has been reported that minimal gyroscope orientation error will cause an acceleration bias of less than 0.01 m/s^2 that will result in a drift of almost 5 m in 30 s of data capture (Welch and Foxlin 2002). Clearly, this is not ideal, and in the mid-2000s, a change in approach to type of data evolved.

Around 2005, people started thinking that inertial sensors be used for what they could capture, i.e. acceleration and angular velocity, rather than taking this data and putting through integration, causing the inherent problems just spoken about. Looking at information such as temporal kinematics in the form of event signals has been validated for some time. Quite high levels of accuracy in gait kinematics such as stride, step, and stance phases were shown possible in 2010 (Lee et al. 2010a). This data was, however, typically labour-intensive to interpret and did require some specialist skills to recognise patterns of movement. This meant that processing times could be less than real time which is not ideal.

1.2 GPS and Other Sensors

Strictly speaking, inertial sensors are just those that measure changes in inertial, namely accelerometers for linear acceleration and gyroscopes (usually rate gyroscopes) for rotational acceleration. However inertial sensors are often combined with other sensors in a single package because of the additional sources of data, when combined or fused with inertial sensor data yield better outcomes (Lee et al. 2012). Two widespread examples of this are GPS and magnetometers.

GPS or a global position system, arguably not a sensor in the strictest sense of the word, uses multiple near earth satellites to determine both position and velocity. Historically, it has widespread applications in navigation systems, but due to trends in miniaturisation and complexity, the technology can now be included in body-worn sensors, as the spatial and temporal resolution is suitable for measuring gross human motion. One of the challenges in using GPS for human locomotion is that it is designed for more steady-state movements of vehicles, rather than the rapid changes in human locomotion. Studies have shown that for this reason its ability to resolve velocity on human athletes can have a substantial error (Wiseby et al. 2010). While its utility to determine spatial position is clear, some researchers have found that velocity can be more accurately predicted using inertial sensors (Neville et al. 2011). GPS position is a most useful measure to include with inertial sensors, because spatial distance (displacement) is an easily understood metric to supplement acceleration data.

GPS devices use comparatively large amounts of power, giving rise to either shorter run times or being larger to accommodate greater batteries. GPS devices require near horizon satellites to give the greatest accuracy to their measures, where these might be obscured such as in built-up areas, indoors or even in large stadiums; this can affect the accuracy dramatically. There was an interesting case in the sport of rugby some years back, where a team that wore in-game GPS devices protested the closing of a stadium's retractable roof in order to avoid loss of GPS data (Neville et al. 2010).

One of the great challenges of inertial sensors is the lack of absolute references, as a zero inertial can exist and any constant velocity. While gravity provides a convenient downward reference source, it can often be occluded by dynamic activity (see later). An alternative reference source is that of the earth's magnetic field, which is in a practical sense constant, for any given geographical region. Magnetometers are available that are sufficiently smaller that today they are often packed together with accelerometers and gyroscopes in inertial monitoring unit (IMU) devices, even though they are not truly inertial sensors. Magnetometers that provide an orientation with respect to the earth's magnetic field are commonly used to determine an orientation "zero" as well as an absolute orientation. This is mostly useful when using inertial sensors for short-term navigation. A good example is limb kinematics like a foot movement over a gait cycle (which can be reset for every heel strike).

1.3 Trends and Availability

Today, inertial sensors are widely available in consumer electronics and specialist movement monitor alike. This has arisen largely through the combination of the technological trends of miniaturisation, mass market adoption and convergence. Trends of miniaturisation saw two specific waves for inertial sensors, first for cars and the second for mobile phones. A second wave also incorporated trends of convergence with wireless communication, computing power, and storage capacity into small, more wearable electronics such as these phones and other wearable devices.

As a result, the popularity of wearable technologies has skyrocketed in recent years with a estimates exceeding $US34B in sales by 2020 (Lambkin 2016). Of these, a considerable portion makes use of inertial sensors, which is the focus of this book. Inertial sensors were not always so popular or accepted; their prominence has arisen because of their size, ability to be integrated with other technologies and availability due to mass production and favourable costs.

Initially, the use of inertial sensors was little more than a spring mass and a switch; in fact, the mechanical pedometer is still quite popular. The first serious wave of miniaturisation came when they were encapsulated into a micro-electromechanical system (MEMS) device, principally for the automobile industry where they were used to detect large changes in acceleration for airbag deployment (Walter 1997). Some years later, they began to appear in smartphones to detect screen orientation, this was accomplished by measuring the direction of acceleration from gravity (down) in the way the tablet or phone, and from there, they found their way into more consumer products such as lifestyle activity monitor such as Fitbit.

The use of inertial sensors requires electronics and low-power computation support systems to enable them to be used. Data must be stored, communicated and analysed and powered by a battery, all in a single wearable device. The continued evolution of computing power has had an enabling effect on the use of these sensors. Moore's law (Schaller 1997) is a widely held maxim in the semiconductor/computing industry which states that the density of transistors (a fundamental building block of computing) will double every 1.5 years or so. This law means also that for a given complexity of technology its size will halve every 1.5 years, so thanks to the laws of compounding, the power of large desktop computers of yesteryear is now available in a smartphone. Also the availability of storage, communication and processing that is growing smaller has also facilitated the adoption of wearables. Miniaturisation has driven a trend called convergence where many technologies that share the same underlying fundamental technologies can now be fit into a single device. Who would have guessed that Nokia (a phone manufacturer) would for a time be the world's largest camera manufacturer and that Apple might lead a watch revolution for a sector in danger of market failure?

Thus, the technologies had a reach from initially high-end automobiles to smartphones, which have now penetrated the market to be the dominant computing device in the most First World countries to widely available wearables that make ideal fashion and convenience purchases. This progressive increase in market size from specialist to widespread has had a dramatic effect on volume and brought downward pressure on costs of production and size as well.

This is terrific news for the tyro looking to try them out in sports applications producing a veritable smorgasbord of technologies to suit the many, many applications. At the time of writing, dedicated medical-grade products such as the ActiGraph and other providers of physiological monitoring are widely used by researchers and clinical practitioners alike. Within the sports sciences, Xsens provides a wearable suit consisting of a network of inertial sensors that is challenging the more conventional motion capture systems for accuracy. On the field, a variety of GPS devices include inertial sensors as well which are widely used in many of the world's most popular team sports—a prominent brand being Catapult, based on some of the authors' early work in the field (Mackintosh et al. 2008). All of these technologies provide time series sensor data as well as derivative measures. Within the consumer-grade electronics, there is a plethora of devices providing often only derivative information, derived from sensors themselves with considerable variation in accuracy. Smartphones and more recently many smart watches also serve as inertial sensor platforms, though the timing accuracy of the samples often has considerable variability (Rowlands et al. 2011).

One of the truly interesting things to consider is how the technologies and wearables sensors we are using today will change based on Moore's law; imagining these possibilities now can help conceive the new applications for research, human movements studies as well as the products of tomorrow. The challenge for the contemporary scientist is stayed current with the latest technology and technological applications. This is surely an issue for future high-performance teams and institutes of sport.

References

R. Herren, A. Sparti, K. Aminian, Y. Schutz, The prediction of speed and incline in outdoor running in humans using accelerometry. Med. Sci. Sports Exerc. **31**(7), 1053–1059 (1999)

D.A. James, N. Davey, T. Rice, An accelerometer based sensor platform for in situ elite athlete performance analysis, in *Sensors, 2004. Proceedings of IEEE* (IEEE, 2004, October), pp. 1373–1376

D.A. James, The application of inertial sensors in elite sports monitoring, in *The Engineering of Sport 6* (Springer, New York, NY, 2006), pp. 289–294

D.A. James, N. Petrone, *Sensors and Wearable Technologies in Sport: Technologies, Trends and Approaches for Implementation* (Springer, Berlin, Germany, 2016)

P. Lamkin, Wearable tech market to be worth $34 billion by 2020. *Forbes* (2016)

J.B. Lee, R.B. Mellifont, B.J. Burkett, The use of a single inertial sensor to identify stride, step, and stance durations of running gait. J. Sci. Med. Sport **13**(2), 270–273 (2010)

J.B. Lee, Y. Ohgi, D.A. James, Sensor fusion: let's enhance the performance of performance enhancement. Proc. Eng. **34**, 795–800 (2012)

C. Mackintosh, D. James, N. Davey, R. Grenfell, K. Zhang, U.S. Patent Application No. 11/843,204 (2008)

J. Neville, A. Wixted, D. Rowlands, D. James, Accelerometers: an underutilized resource in sports monitoring, in *2010 Sixth International Conference on Intelligent Sensors, Sensor Networks and Information Processing (ISSNIP)* (IEEE, 2010, December), pp. 287–290

J. Neville, D. Rowlands, A. Wixted, D. James, Determining over ground running speed using inertial sensors. Proc. Eng. **13**, 487–492 (2011)

D. Rowlands, D. James, Real time data streaming from smart phones. Proc. Eng. **13**, 464–469 (2011)

R.R. Schaller, Moore's law: past, present and future. Spec. IEEE **34**(6), 52–59 (1997)

P.L. Walter, The history of the accelerometer. Sound Vib. **31**(3), 16–23 (1997)

G. Welch, E. Foxlin, Motion tracking: No silver bullet, but a respectable arsenal. IEEE Comput. Graphics Appl. **22**(6), 24–38 (2002)

R. Williamson, B.J. Andrews, Detecting absolute human knee angle and angular velocity using accelerometers and rate gyroscopes. Med. Biol. Eng. Compu. **39**(3), 294–302 (2001)

B. Wisbey, P.G. Montgomery, D.B. Pyne, B. Rattray, Quantifying movement demands of AFL football using GPS tracking. J. Sci. Med. Sport **13**(5), 531–536 (2010)

T.C. Wong, J.G. Webster, H.J. Montoye, R. Washburn, Portable accelerometer device for measuring human energy expenditure. IEEE Trans. Biomed. Eng. **6**, 467–471 (1981)

Chapter 2
Theory and Application

In this section, we examine the types of signals from inertial sensors, in particular, the accelerometer using the basics of sampling and digital theory. An understanding of these theories is essential to utilise these sensors in developing applications for human motion analysis. What we are going to be demonstrating here (together with the limitations and advantages of operating in a digital world), are all the impact design and experimental procedures to get the best from this technology. Following this we examine the equations of motion for inertial devices, using examples from the literature to get a better understanding of the signals we are hoping to extract meaningful data from.

2.1 Digitisation

The use of inertial sensors in the present day uses almost exclusively digitally converted samples. It is therefore important to have a basic idea of digital processes to ensure an understanding of the advantages and limitations (Cutmore et al. 2007). These considerations will lead to more robust experimental designs and treatment of data that has been collected using these types of technologies.

Any moving segment of a human body, e.g. a thigh, is undergoing continuous changes to its inertia, which can be understood through changes in the linear accelerations and angular accelerations (rotations) of that segment that an inertial sensor is attached too. While we have, in principle, a continuous data source from movement that any sensor can detect, digital based recordings give rise to discrete measures (how often we sample) and accuracy of measurements (the resolution of the measure). Key concerns are then that we need to ensure we are sampling at often enough through a sufficiently high sample rate and with enough accuracy to resolve the measures for our intended application. The case examples in the final section will give examples of how the understanding of this codependency is

critical for individual applications. First, though an understanding of the definitions
of sample rate, resolution and range are needed.

2.1.1   Sample Rate

Figure 2.1 shows a continuous waveform (dashed line) together with discretely
sampled data (circles). Samples are usually taken over time, with sample time
referring to the time between same and the sample rate the number of samples in a
second. For example, a 100 Hz sample rate has a sample time (or period) of 0.01 s,
with 100 samples per second as the sample rate. From this figure, we can deduce
that reconstructed sampled signals (straight line) are not exactly the same as a
continuous signal instead they are an approximation. The challenge is then to
determine what is a close enough approximation for the sampled signal to accurately represent the signal of interest. In a perfect world, we could sample infinitely
fast and with infinite accuracy; however, this would lead to an infinite data size and
thus is not practical given the limits of data storage and processing of the data. What
then is the effect of accuracy through the resolution of the signal.

2.1.2   Resolution

In addition to the sample rate, the resolution of the signal measured is another
important consideration. Because most sampling done today is digital, the resolution
is often expressed in bits. Bits, with a value of only one or zero, are the fundamental
building blocks of the digital world. As we increase the number of bits available for
storage, simultaneously there is an increase in the number of combinations of ones or
zeros a sampled signal can be stored into. Note: such combinations are typically
known as discrete values. Thus for 2-bit storage, four possible values can be resolved

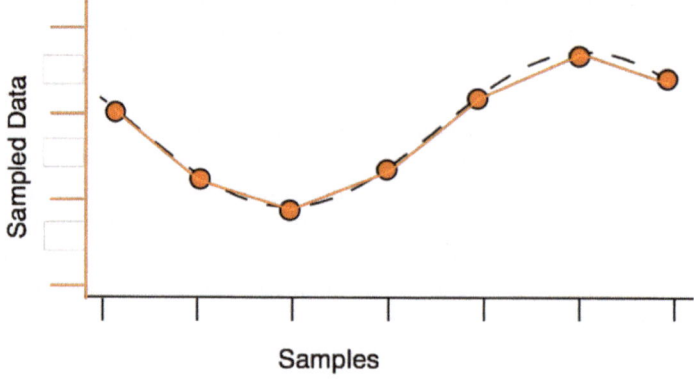

Fig. 2.1 Digitisation of a continuous waveform

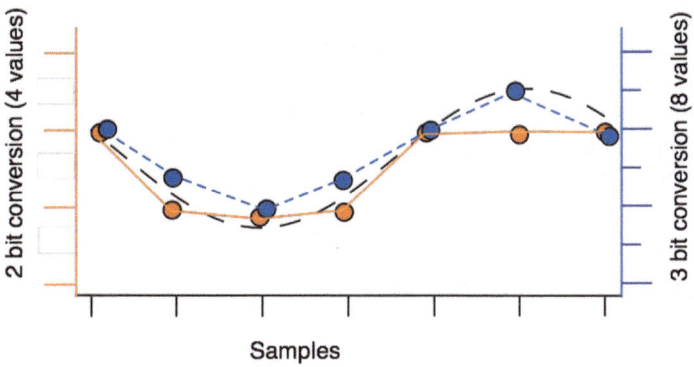

Fig. 2.2 Digitisation of a continuous waveform with samples at two different resolutions

(00, 01, 10, 11) or 2^2. Commonly today resolutions are often 10 bit, 16 bit or more. More bits thus give a finer resolution of what is being measured. Figure 2.2 shows our example signal digitised into a 2-bit (solid line) and 3-bit signal (small dashed line). You can see here the higher the resolution, the more closely the reconstructed signal matches that of the original signal (dashed line)

Putting everything together in Fig. 2.3, it is clear that a higher sample rate and resolution (solid line) offers a superior representation of the original signal (dashed line) than that of a lower sample rate and resolution (fine dashed line). It would be an easy yet simplistic conclusion that the higher the sample and resolution the better. This however has a knock on effect, specifically on memory and battery capacity to run such devices, so it is important that some understanding of what is to be measured and the subsequent when making considerations for sample rate and resolution.

For human motion, a typical rule of thumb is that you need to sample an order of magnitude or ($10\times$) the signal of interest. For example, if stride rate is 2 per second (2 Hz), then you would need to sample the signal 20 times per second (20 Hz). In the world of signal processing, all signals can be represented as a series of sine waves of different frequencies. These are sometimes called frequency analysis or

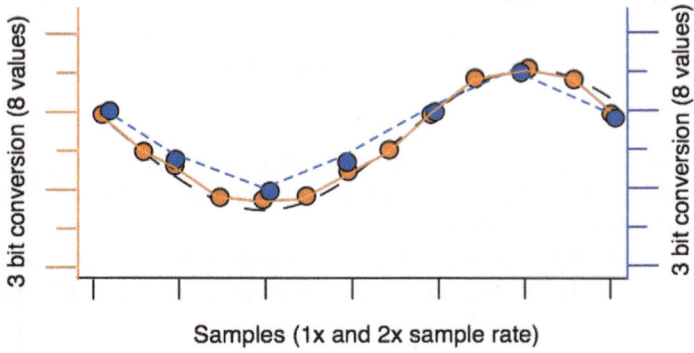

Fig. 2.3 Digitisation of a continuous waveform at two different sample rates

spectrogram or with the fast Fourier transform (FFT) being the usual technique used. While this is beyond the scope of this work, it is worth bearing in mind if you want to understand your data a bit more.

An important outcome of this analysis is the Nyquist frequency that says you need to sample at least twice as fast as the highest component of frequency you are interested in a signal.

2.1.3 Resolution and Range

Where a range of numbers must be squeezed into a set number of unique digital combinations. If we wanted to store values of acceleration from $0 \to 1$ g's with a 4-bit digital storage, we would have 16 unique values (2^4); thus, each digital value would correspond to 1/16 g or 0.0625. In example, the range is 1 g and the resolution is 0.0625 or 62.5 mg.

In most inertial sensors, today the number of bits is around 16 bits (though 8, 10, 12 are quite common too) and the range is often selectable from ± 2 to ± 1 16 g. As the range is usually expressed as +/1 g, your range is actually centred around 0 g. In such a case, if you are wanting to measure accelerations up to say 10 g, sensor would also likely measure down to -10 g for a total range of 20 g.

Thus, range and resolution are related by the following equation

Range $=$ highest $-$ lowest values that the sensor can be detected (Most sensors today are
           equally spread around zero)

$$\text{Resolution} = \text{Range}/2^n; \quad \text{where } n = \text{number of bits.}$$

Data storage for a sensor is simply the product of the number of bits for conversion x number of channels x sample rate. Thus for 100 Hz sample rate using a 10-bit converter with 3 channels (x, y, z), you will be storing 22 kilobytes per minute. While this may not seem a lot, if you are wanting to use the sensors for extended periods, or have much higher accuracy (bits), or want to use a lot of sensors for multiple body segments this can quickly become challenging and have downstream effects on areas such as: storage, battery usage, or data transmission challenges.

Understanding the range of physical activity you are interested in measuring determines the sensor you select for your application. Understanding how finely you want to discern movement determines the number of bits you will need to acquire (and store) for subsequent analysis.

It is a common trap to think that the highest range sensors are best for any given situation; for example a ± 50 g sensor has a corresponding range of 100 g and with an 8 bit converter you would only have a resolution of 0.39 g (where 1 g = earths normal gravity). Equally if you decide to use a 32-bit converter to determine body roll of a swimmer (1 g range) using a ±2 g sensor, your resolution of 1 ng will certainly be accurate enough, but your storage requirements will be large.

2.1.4 Calibration

Using sensors whether off the shelf or built into commercial devices, it is critical to have a basic understanding of calibration. The fundamental devices themselves are quite robust with low noise, high linearity and low drift rates (Skog 2006). However downstream, due to being connected to support electronics and are processed digitally, inevitability some uncertainty can creep in and this has an effect on the interpretation of the data collected and its subsequent utility. For this reason, an understanding of calibration is important.

If we take an assumption of linearity, then it remains the task to categorise the linear variables for each sensor in use. In the case of accelerometers, gravity (9.8 m/s/s) makes an ideal reference source, and by turning a sensor 180°, its opposite makes a second point for calibration. This is probably the simplest method to calibrate sensors, assuming this is the prime magnitudes of interest (which is usually the case for human locomotion). Thus, recording data and placing the sensor on each of its three axes, positive and negative with respect to gravity, make a convenient procedure before undertaking any recording. It serves also to check channel corresponds to each axis of interest.

Of course more sophisticated methods can be used to calibrate such as the Lai method (Lai 2004). Figure 2.4 shows a three-axis accelerometer calibration trace using gravity as a reference. By reading the average values of each square wave, a zero offset and number of counts corresponding to gravity can easily be read off.

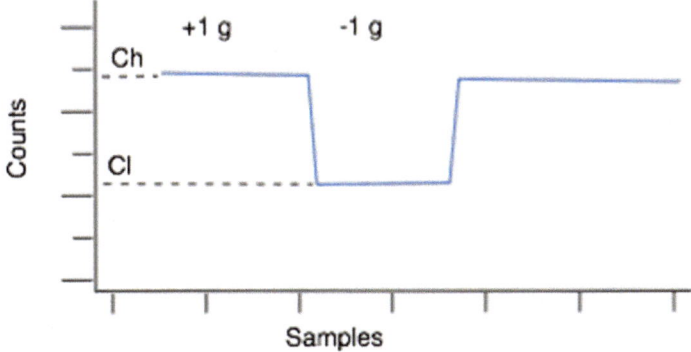

Fig. 2.4 Calibration of an accelerometer using gravity

For gyroscopes, audio turntables make a convenient method of checking angular rate in each of the corresponding channels.

References

T.R. Cutmore, D.A. James, Sensors and sensor systems for psychophysiological monitoring: a review of current trends. J. Psychophysiol. **21**(1), 51–71 (2007)

A. Lai, D. A. James, J. P. Hayes, E. C. Harvey, Semi-automatic calibration technique using six inertial frames of reference. In Microelectronics: Design, Technology, and Packaging **5274**, 531–543. Int. Soc. Opt. Photonics. (2004, March)

I. Skog, P. Händel, Calibration of a MEMS inertial measurement unit. In XVII IMEKO world congress 1–6 (2006, September)

Chapter 3
Acceleration Components

Newton's laws of motion (presented here for straight line motion) show the relationship between displacement (distance), velocity (speed) and acceleration. This demonstrates velocity over time produces distance and acceleration over time changes velocity. Therefore, distance travelled can be calculated from acceleration.

Motion equations show an initial velocity and a constant acceleration; however, for human motion analysis, there are no constants. Acceleration itself is determined by a multitude of components which we will investigate in this section.

The challenge in measuring acceleration (linear or rotational) is that it is two steps (a double derivative in fact) of that with which our mind understands physically, that is distance. In relating inertial sensors to our usual 3D physical environment and Cartesian coordinates we must adjust our thinking from what we can see (position), infer velocity (changes in position over time) and acceleration (which cannot be seen). Having a firm grasp that acceleration data is entirely different (though related) to positional data is critical to beginning to use it to gain meaningful and often complementary insights. A simple example of this is looking at heel strike using video (or motion capture) compared to an accelerometer (or a force place for that matter). Ground contact occurs before force transfer (and a change of acceleration).

The resulting acceleration experience on a body can be expressed as the sum of four components (Ohta 2005). These components are the inertial changes of the body, gravity, centrifugal motion and tangential acceleration. Each of these combines to the overall acceleration that will be measured by the sensor. In seeking to use sensors for human locomotion, it is essential to understand the role each of these plays in individual cases to effectively decouple, remove or make best use of each for the interpretation of data. The differences are manifested in differing magnitudes, temporal or frequency domain enabling modern filtering techniques (often quite simple to implement) to separate the components.

Note: Introduce some text about inertial frame of reference concept somewhere. This is where velocity is constant—usually zero in laboratory equipment. Sensors work in own local coordinate geometry that is constantly changing orientation—this is the mental challenge to decouple.

3.1   Inertial Component

These are the direct linear component accelerations measured by a sensor. In practice, these components are often much smaller than the other components, though a few exceptions make it one of the most practical of the components for understanding and decoupling human motion. Rapid motions of limb segments make a contribution as do rapid periods of decelerations such as contact events (100's of g's) for measuring percussive forces such as striking in boxing and ground contact events in running and walking gait.

In the case of rapidly moving limbs, it is important to remember that accelerations are not velocity and not displacement; thus, peak acceleration in a movement is likely to be at the start acceleration and end deceleration of a movement, opposite in sign and hovering around zero in the middle of a movement. By contrast, the velocity will be at its maximum in the middle of the movement and displace at its highest at the end of a movement.

For percussive events, deceleration happens during contact, and depending on the shock attenuation such as padding on a boxing glove, the time for the contact event can be very short, yet very high in magnitude. In such cases, sensor will usually go over range and the samples may not catch enough points on the waveform to represent the event accuracy. They do, however, make for very good temporal markers of the event itself—something which aids further analysis especially when coupled with other sensors. This will be explored further in the case studies.

3.2   Gravity Component

The earth's gravitational forces create in a practical sense a constant form of acceleration in a given inertial frame of reference. Any changes to this value are then the results of changing forces on the sensor itself. Thus, understanding that constant value should be present enables the analysis of changes at the sensor position, the most common of which is a change of orientation of a sensor and the location of the vertical axis. Depending on the form of locomotion, this takes of different forms and can be more easily understood through a few examples. In Fig. 3.1, a sensor is attached to the sacrum (small of the back) of a swimmer. As the torso rolls during swimming action, the gravity component is seen alternately in the sagittal and frontal axis of the transverse plane. It thus forms a convenient measure to count swimming strokes, ultimately leading to measures of stroke phase and lap turn times as well (but more on that in the case example later on).

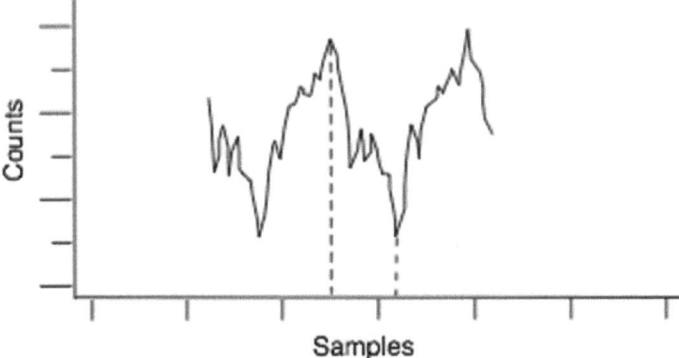

Fig. 3.1 Gravity component from swimming strokes on sacral mounted sensor

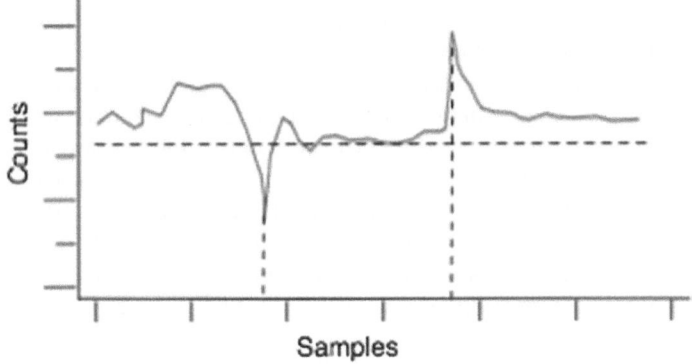

Fig. 3.2 Flight time from takeoff to landing, determined by the acceleration peaks

For human movement where orientation is important to determine a particular activity, detection of the gravity component can be used to classify limb segment orientation to thus determine activity type. Examples such as daily living tasks for health applications (James et al. 2012) and tackling in ball sports (Delaney et al. 2016) show comparable levels of accuracy to gold standards like video or motion capture systems. The chief advantage here is that these can be computationally derived and used in the ambulatory environment (Gleadhill et al. 2016).

Often, it happens that during human locomotion, the body and attached sensor are no longer in an inertial frame, such as in jumping motion. In this case, the gravity component does not appear in any of the channels and can serve as a reliable measure of time of flight (Fig. 3.2) and ultimately jump height. This has been applied to team sports such as volleyball (Gageler et al. 2015) and snowboarding (Harding et al. 2007). In the latter example is was validated as being as accurate as high speed video and being of much greater convenience. Additionally,

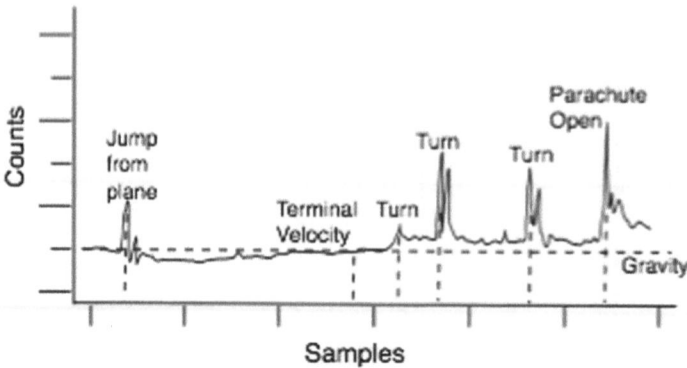

Fig. 3.3 Skydiver acceleration traces from body-worn sensors

the sensor output in the snowboarding study was found to have a high degree of correlation with more subjective measures, such as judging an athlete's performance during actual competition.

The gravity component transitioning from a non-inertial and inertial frame of reference can be seen clearly in the figure below of a skydiver undertaking acrobatic manoeuvres below. In this figure, the skydiver is in free fall (no gravity component is present) until terminal velocity is reached (gravity component can be clearly seen) and then as the skydiver undertakes acrobatic tumbles the component transitions between the various axes clearly delineate their activities. This work (Wixted et al. 2011) was envisaged as an alternative method to judge the spot to the traditional use of a telescope from a distance of ~ 2 km away (Fig. 3.3).

3.3   Centripedal

Where a moving body, body segment or sensor on such is moving in a constrained radial direction, a centripetal force is exerted towards the centre of rotation. Sometimes, this is expressed as an outwards force called the centrifugal force. Often, the correctness of the terminology is debated among the various scientific disciplines. The focus here, however, is that this additional acceleration component is measured by a sensor, as a component of the inertial acceleration equation presented earlier. This component can be found in low magnitudes in arm, and leg swing events during ambulation, however, can be somewhat swamped by ground contacts of other limb movement components such as throwing. In the case of throwing, or bowling in the sport of cricket, the centrifugal component can become the most dominant of components.

In a study examining bowling action in cricket (Spratford et al. 2015), where a straight arm is used to accelerate a ball before release, rather than articulating the elbow for the more conventional bowling motion, the centrifugal components can exceed 50 g and the degree of phase correlation between the upper and lower arm can accurately assess whether the ball has been bowled or thrown to assess its adherence to the game.

An important consideration in this study was the dynamic range of the sensor (±50 g) meant that it had little resolution available for detection of running activity (±2 g), and so a dual sensor was required for this application.

References

J.A. Delaney, G.M. Duthie, H.R. Thornton, T.J. Scott, D. Gay, B.J. Dascombe, Acceleration-based running intensities of professional rugby league match play. Int. J. Sports Physiol. Perform. **11**(6), 802–809 (2016)

H.W. Gageler, S. Wearing, A.D. James, Automatic jump detection method for athlete monitoring and performance in volleyball. Int. J. Perform. Anal. Sport **15**(1), 284–296 (2015)

S. Gleadhill, J.B. Lee, D. James, The development and validation of using inertial sensors to monitor postural change in resistance exercise. J. Biomech. **49**(7), 1259–1263 (2016)

J.W. Harding, K. Toohey, D.T. Martin, C. Mackintosh, A.M. Lindh, D.A. James, Automated inertial feedback for half-pipe snowboard competition and the community perception. The impact of technology on sport II, **2**, 845–850 (2007)

D.A. James, R. Leadbetter, N. MadhusudanRao, B. Burkett, D. Thiel, J. Lee, An integrated swimming monitoring system for the biomechanical analysis of swimming strokes. Sports Technol. **4**(3–4), 141–150 (2012). https://doi.org/10.1080/19346182.2012.725410

K. Ohta, Y. Ohgi, H. Kimura, N. Hirotsu, *Sports Data* (Kyoritsu, Tokyo, Japan, 2005)

W. Spratford, M. Portus, A. Wixted, R. Leadbetter, D.A. James, Peak outward acceleration and ball release in cricket. J. Sports Sci. **33**(7), 754–760 (2015)

A. Wixted, D. James, Inertial monitoring of style & accuracy at 10,000 feet. Proc. Eng. **13**, 493–500 (2011)

Chapter 4
Case Studies

As with any scientific-based research, there is a procedure that should be followed. It is known as the scientific method (Fig. 4.1). This tried and true method is followed to answer a question that a scientist may have. This is often where an observation has been made, e.g. a deficiency in technique that may affect sporting performance. Therefore, a design for an intervention needs to be implemented. Part of this process is to choose the most suitable technology to carry out an assessment. To determine the technology to be used, the researcher needs to know the precision and accuracy of the chosen device. For new products, this is typical, and this is achieved by validating the technology against an accepted method. Traditionally, Bland–Altman limits of agreement have been used (Bland and Altman 1986). More recently, the Will Hopkins typical error of the estimate is becoming increasingly popular in sport science-based validations (Hopkins 2015).

4.1 Approaches

4.1.1 Study Design

A validation of new technology or method should be made by comparing it to an accepted or established equivalent for the particular application. While it cannot be always possible, the criterion to be measured against should be ideally the gold standard. If agreement analysis is made, the stronger the agreement between the new technology and gold standard, the more confident a user can be of its accuracy and precision. With an estimate of the error, the ideal is for trivial error. Validation is specific. For example, if wearable sensors were validated for freestyle swimming stroke rate, a further validation would be required for stroke rate for each of the three form strokes, i.e. butterfly, breaststroke and backstroke. Once validation has been accepted, sports application is possible.

© The Author(s), under exclusive license to Springer Nature Singapore Pte Ltd. 2019
J. Lee et al., *Wearable Sensors in Sport*, SpringerBriefs in Applied Sciences and Technology, https://doi.org/10.1007/978-981-13-3777-2_4

Fig. 4.1 Flow chart
demonstrating the processes
that make up the scientific
method

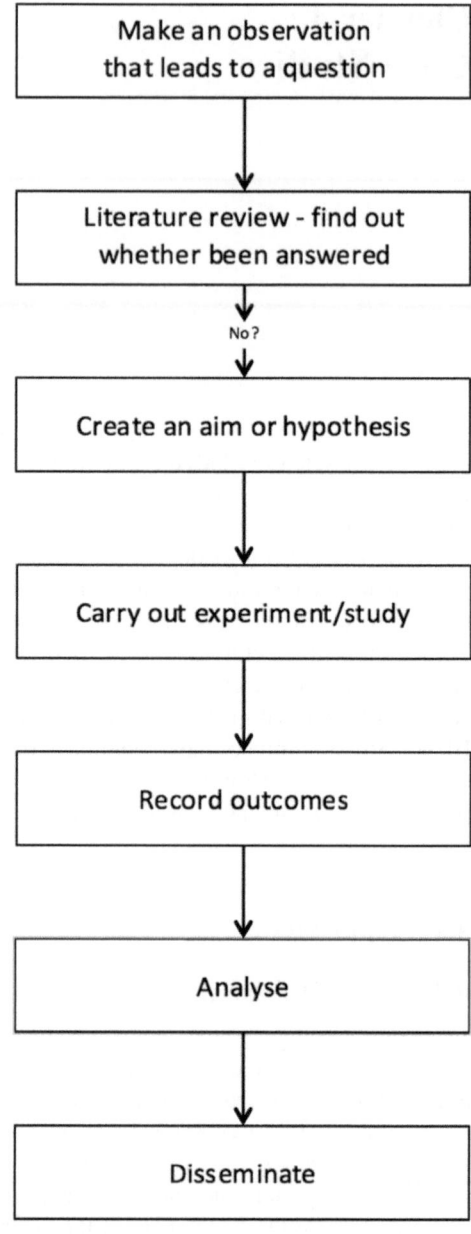

Any research that involves humans as participants typically requires clearance for ethical research by an institution's Human Research Ethics Committee (HREC). National bodies develop ethical guidelines to follow. These guidelines are based on what is known as the Helsinki Declaration from the 1960s, and its ongoing intention is to protect human rights in medical and related human experimental trials. There are very few occasions where an application to an HREC is not required. Some pilot studies may not need approval. However, at the very least, researchers should liaise with their institution's HREC to determine the need to apply for ethical research approval. Until an approval is given, no research should commence. Especially for long-term projects, there may be reporting requirements to the relative HREC.

4.1.2 Recruitment

Recruitment of participants is very important. Those that express an interest must be reflective of the population that the research is intended to study. For example, a novice swimmer may not be suitable in an early swimming study if the aim was to develop a model swimming profile that swimmers may wish to emulate. Therefore, elite athletes should be the preferred option for this purpose. It can be difficult to attract people to participate. Once volunteers are present and ready to participate, the process should be well planned and executed. Otherwise, the volunteer's valuable time is wasted and that often results in participant dropout. Information sessions are the effective ways to inform people of the research. A common alternative is the use of a plain language statement (PLS). This is typically designed to provide thorough information that is easy to understand by the layperson. A PLS goes hand in hand with a consent form. Once a potential participant expresses an interest to be involved, they need to complete and sign a consent form. A good consent form and PLS will provide enough information to a potential participant (or guardian if required) to be able to make a decision whether to volunteer or not. The basis of the information should be whether the benefits of the research outweigh any risk imposed on participation. Clearly, the risk is towards the participant, while the benefits made not directly flow back to them. Therefore it must be clear what are the risks to the person and who may benefit from them participating. Any risks and benefits, along with the PLS and consent forms, should have been approved during the HREC application.

4.1.3 Technology Pilot

The application of technology to assess human movement has enabled the performance enhancement of athletes. The continual development of technology continues to grow the range of choice for a sports scientist or coach to utilise. With each new product made available, the choice grows. This can cause a paradox. The suite of available technology may result in either confusion on what to choose or a user is overwhelmed to the point of being unable to know what to choose. This leads to an important consideration for technology developers: usability.

Understanding or knowing how to find out what the end-user wants is at worst not considered or at least assumed what is wanted or that users will easily "pick up" how to use a device.

From a development point of view, once a piece of technology is developed and before a participant-related pilot study is carried out, a technology pilot should be implemented. This can include bench-testing the device. This will provide performance levels of the technology and give an indication of any parameters that the device may be restricted too. Therefore, it can be determined whether the technology is suitable for a particular type of data collection, e.g. accelerometers to measure body roll in freestyle swimming. Once the technical piloting satisfies the researcher, participant-included pilot studies can be made.

4.1.4   Pilot

There are many issues within the human factor component of technology development. For an engineer to properly understand, the range of factors will provide them with a skill set that may enhance greater use of their technology through a device "fitting" a need. Not only technical selection has to be made. There are other considerations to be carried out.

A major consideration is the design of a pilot study. An early objective of a pilot study is to sort problems and ensure the data collection procedure works effectively and efficiently. If a pilot is bypassed, the researcher runs the risk of volunteers to the project being unnecessarily messed around while problems are sorted. The risk of problems is always present; however, these can be mitigated somewhat with an effective pilot to trial the data collection processes. Choosing a suitable facility is important to take into account. For example, if a study was to look at tumble turns during 1 km of non-stop swimming, it would possibly be better to choose a 25 m pool in preference to a long-course 50-m pool. This would allow for twice as many turns to be assessed, i.e. 40 instead of 20. Equally what occurs in still open water could not be used to make assumptions to conditions where the surf is encountered.

4.1.5   Full Study

When researchers are confident that any problems have been shown from the pilot study, a full research typically follows not long after. By this stage, ethical clearance should have been received. If not, until clearance is given, the research cannot proceed. The first step is recruitment mentioned a few paragraphs back. Scheduling of participants is very important, and while the pilot will have provided a big indicator of how long a single data capture will take, it is usually wise to allow plenty of time at first. It is almost invariable some hidden problem will crop up. This may be as simple as overlooking the need to provide directions to the place for data capture! This is where project management skills assist a researcher being able to run data capture smoothly. Again with the pilot study, it should be apparent how to manage data once collected. For example, can it be collated as it is collected? Or will the data need to be processed in some way before storage. File management is also very important too.

Inconsistent filing will almost invariably result in major problems later. This is especially the case when the data is captured over long periods of time and if the volunteers return for multiple captures, e.g. pre- and a post-exercise intervention trial. Ethics requires data to be de-identified and at this first point is ideal. In the case of validation type research, it is usually single visit data capture, and there is no need to create a file where the de-identified data can be re-identified or in the case of test/retest research. In such cases, a separate file is usually needed where participant names are kept and associated with a code. That code usually goes into data filenames. A date is very handy in a filename too, as is a trial number if retesting is part of the research. All these procedures will need to be explained in the ethics application. This is also the case for details such as participant numbers, the research (data collection) protocols, among much else for ethical research. Any deviation from the approved procedure will need a variation to the original approval. For this reason, along with being able to recall what occurred, time taken, participant feedback, etc., all details need to be recorded and kept. These being kept and constantly referred and added to during that capture enables the researcher to make sure they keep the data collection consistent and in line with all requirements.

4.1.6 Analysis and Interpretation

Once data has been collected, there are several processes required. This includes signal processing, filtering and trimming to manageable and relevant files. Appropriate statistical analysis has to be chosen. This is all followed by reporting of the results. When statistics are used, the outcomes give the researchers an understanding of how confident they can be of making a prediction or claiming a relationship between the variables. A common test for probability is a T-test. There are several iterations of T-tests; however, they all measure the probability of what is being tested occurring. It is worth pointing out here that T-tests are reported as "p" values because of the "probability" and correlations are reported as "r" values because of the "relationship". Typically in a T-test, the level of significant difference (often termed as alpha level) is quite often set at $p = 0.05$, which in basic terms there is a 95% chance of the data *not* being significantly different; e.g., swimming breathing strokes in a pool are different compared to those in still open water. However, this can be changed, and the more critical the analysis, the lower the alpha level should be. Meaning that $p = 0.01$ shows that to have a significant difference, 1% of the data will be different.

During the analysis process, decisions are typically made as to the optimal way to present data. Generally, the choices include graphical, table or written formats. What is important is that the data is presented in a logical and understandable format. A Results section is arguably the most difficult section of a report to follow. Clear representations that can be followed should be made. Hence the reason for careful consideration in order to make the correct decision of how to best present the results. One strategy to see whether a figure or a table is understandable is to have someone who does not know the objectives to take a look at it, and with minimal prebriefing, see if they can describe what it means. If the person struggles, the item they are looking at is likely it will not be understood in a report.

After the results are collated, interpretation is carried out. This is typically reported in the discussion section of a report. A discussion section usually starts with a short paragraph giving an overview of the research, and often a repeat of the research aims is included. This is followed by a series of paragraphs where each contains one of the outcomes that are reported in the results section. Each paragraph should be constructed in the following order:

- A sentence referring to one of the reported results: for example, there was a significant difference between left and right body roll in the first lap swum.
- A sentence on how this relates to previous research: for example, this supports previous research that reported asymmetries exist in swimming styles (Smith and Smith 2018) *or* if the outcomes differ; e.g., this outcome is different to previous research where Jones and Jones (2018) reported consistent body roll actions during freestyle swimming.
- This is followed by a short discussion that interprets why the outcomes in research always appear consistent (NOTE: this is not always the case) *or* why the current research differs to previous research. The paragraph should not ramble on and needs to stay on topic, i.e. the results outcome stated in the paragraph's first sentence.

A new paragraph commences with another reported outcome from the results. Following all the reported results and related discussion, a conclusion summarises the research, outcomes and interpretations.

The following sections take a case study approach to demonstrate the effectiveness of wearable technologies to assess human movement. The two chosen examples are gait and swimming. These could equally apply to any function or activity that involves the need to measure movement. It is that some points in the sections that follow can be seen as common between the cases, whereas there are differences present too. Therefore, readers should develop an appreciation of the similarities that span many and possibly all applications. Equally, acknowledge and, if considering developing their own technology, understand that there are differences and not to assume if the technology works one way for one application, it will work for anything—because in all likelihood it will not.

4.2   Gait (Walking/Running)

Gait is the means to move from one point to another via a cyclic kinematic input to create motion. For human movement, gait is almost exclusively inferred towards walking or running. For both walking and running, gait has often been referred to as a series of controlled falls (Novacheck 1998). This is due to the body's centre of mass (CoM) which is positioned forward and outside the base of support, i.e. the external parameters of the feet.

Gait is arguably the single most important activity that involves movement of the human body. Without being able to move about, little else is achievable. Therefore, it should be no small surprise that gait is a highly studied area in human movement research. Prior to research, several points should be taken into consideration.

First, what is the question that has arisen from an observation? This will determine what the research is. In the case of technology use, if it has not already occurred, has the technology been validated to assess gait. Not only that, has it been validated to measure gait specifically for a particular kinetic or kinematic output? If it is kinetic measures, the ideal criterion to compare too is force platforms. Depending on kinematic measures will determine the type of device to measure. Relative to gait and the use of wearable devices, often kinematics of gait events are often looked at. Therefore, camera systems, both 2D and 3D, are common criterions.

Apart from methods of data capture, there are many other considerations to be made, bot prior and post-data capture. As mentioned in Sect. 2.1, the capture frequency is very important to apply correctly. The environment where the capture is taking place has to be taken into account; e.g., is it in a lab or a real life situation. This will determine that capture capabilities such as capture volume; e.g., can only a few step be samples or can the capture last for extended periods. This will lead to data size issues. Large datasets over multiple channels and/or devices are often a challenge. Therefore, time considerations of the capture should be made; e.g., is it worthwhile capturing at 250 Hz for 30 min when a 10 min sample at 100 Hz will be adequate? Oversampling can be very time-consuming during capture, post-signal processing and subsequent analysis. While it is good to capture as much as possible, a "hit and miss" approach can make the research unmanageable. There have been publications that report calculations to determine appropriate capture frequencies such as the work produced by Davey et al. (2007).

Post-capture processing can include data collation and processing into usable files. The application of a filter is often used to clean up unwanted noise. For gait, a 6 Hz low-pass filter is common and occasionally a moving average is preferred. The typical low-pass filters are either Butterworth or Hamming that are chosen. Once processing has been completed, analysis is undertaken. Often with new technology or processes, a validation study has to be carried out. This is aimed at demonstrating how much agreement the output is compared to the true value that was measured. This is often carried out by measuring the exact same activity with an accepted mode of measure for that activity (most preferably the gold standard)

(a)   **(b)**

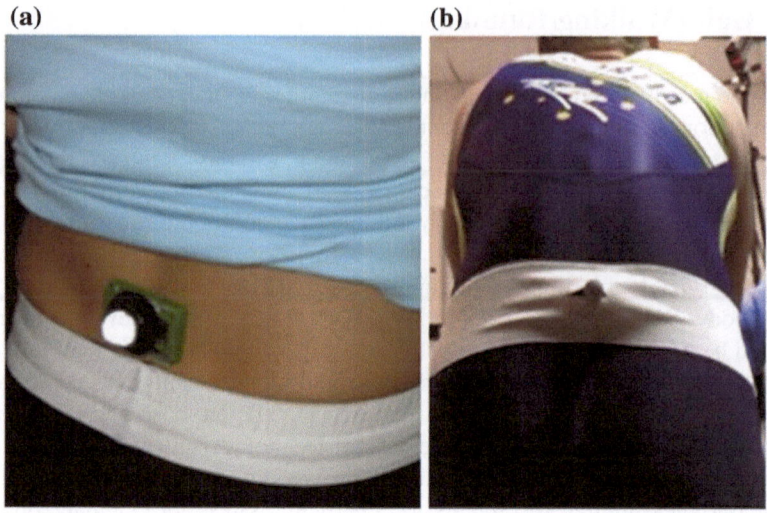

Fig. 4.2 Validation of movement measures of inertial sensor acceleration data with the sensor place at the L5/S1 region of the lower spine: **a** mounted directly on the skin and **b** mounted on the running suit of the volunteer. The reflective marker over the sensor is from an infrared camera system. Movement data from both sources were synchronised for later validation comparisons

and the technology that is being tested (Fig. 4.2). What is being determined is how much the two different methods agree. The closer the measures, the more the agreement. Depending on what is to be measured will determine the criterion (gold standard) measure. For gait, this can be infrared motion capture, force platforms, high speed cameras which are all often used in controlled laboratory settings. The long accepted agreement measure has been the Bland–Altman limits of agreement (Bland and Altman 1986). This was originally designed for clinical validations and often relies on reasonable sized sample groups. A recent method has steadily become accepted in human movement and new technology validation. This is known as the typical error of the estimate and developed by Hopkins (2015). The limits of agreement look at the spread of the data. Typically, this is 95%, meaning that there are a upper and lower limits that are set by 95% of the data. The narrower or closer the limits are to the mean, the greater the agreement between the two methods of measure.

There are many factors that contribute to determining what sample and its size should or can be used. Just like over- or undersampling when considering the time and capture frequency, sample size of data can equally be excessive or inadequate. Too large a sample results in time wastage, especially if the research involves volunteers who freely give up their time to assist in the study. Too small a sample makes, it less confident that any inferences being made are accurate of the whole population that the sample represents. Often, an estimation and justification of sample sizes have to be made early in the design of a research project. This can make up part of an ethics application to carry out research. Other factors that influence a sample are population numbers. In some sports, especially at the elite level, there are at times very small numbers of athletes that compete. As a rule, the larger the standard deviation in the data, the larger the sample size needs to be. This means that the smaller the standard deviation, the more confident that any trend that is seen in a tested sample will be also seen in any other randomly chosen sample of the same population.

For this case study, wearable technology was used to detect heel strike and whether it varied between walking and running. The sensor was positioned on the lower back at the L5 S1 vertebral region (Fig. 4.2). This position is the closest external point on the human body to the centre of mass. With a triaxial accelerometer, it can measure acceleration in the anteroposterior (forwards/backwards), mediolateral (side to side) and vertical directions. This is the lowest position on the body that gait kinematics can be detected from both left and right feet from one point on the body. Therefore, a single device can be used to detect gait events such as heel strike (initial ground contact) and toe off (loss of ground contact) during walking or running.

A heel strike gait event is typically easiest seen in anteroposterior (direction of travel) data (Lee et al. 2010a). If a moving average filter is applied to the data, it typically retains the important events of heel strike and toe of, which are the important identifiers of stride, step and stance. However, the vertical acceleration looks somewhat synodal. This provides an opportunity to use the vertical acceleration as a reference point of where gait events occur (Fig. 4.3). Walking shows heel strike occurring on the right side of the vertical acceleration peak and running on the left. This is possibly explained by the differences in walking where the vertical displacement of the centre of mass is highest at midstance in walking. This is known as the inverted pendulum model (Winter 1995). As a point of important delineation between walking and running, the CoM is at its lowest at midstance during running and the explanation is described as the spring mass model (Blickhan 1989).

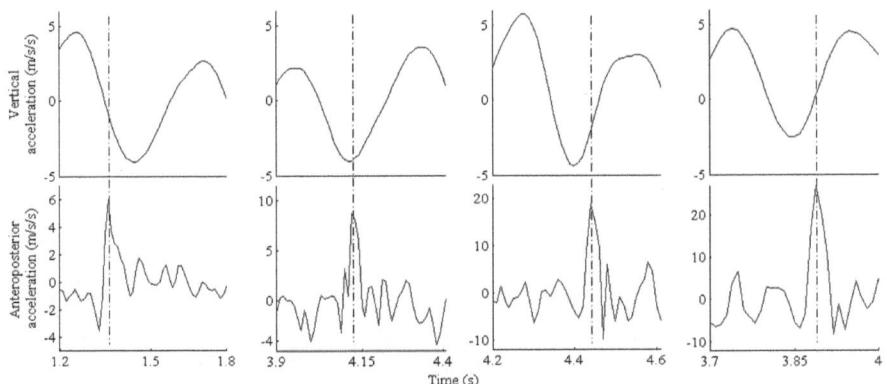

Fig. 4.3 Comparison of heel strike events for walking and running

Being able to identify gait events that enables determination of whether the person is walking or running is quite important. This offers a wide scope for the use of the data. For example, knowing whether someone is walking or running and for how long means that estimation of physical activity being performed. This can then be applied in areas for exercise intensity and periods of time in certain training zones. It can also be used for people in a more clinical sense, e.g. weight loss. By knowing whether someone is walking, running or even doing nothing gives a professional the chance to monitor long periods and determine whether their client's activity levels are sufficient enough to cause weight loss.

Other benefits for positioning a sensor on the lower back are that gait symmetry is measurable from a single device. The mediolateral data can be used to identify gait events occurring on the left or right side of the body (Fig. 4.4). This can result in data being interpreted for walking or running symmetry. From this, several possible monitoring situations can be applied. Changes in running speed of 1 km/h have been shown to affect vertical acceleration data (Lee et al 2010b). One such situation would be monitoring people during rehabilitation after injury, e.g. a sprained ankle. The running kinematics would most likely be affected from an injury of any sort. Therefore, tracking the rehab process is possible. This would be especially viable if a baseline measure could be used as a reference. For example, at the commencement of a season, an athlete has a baseline profile of their running

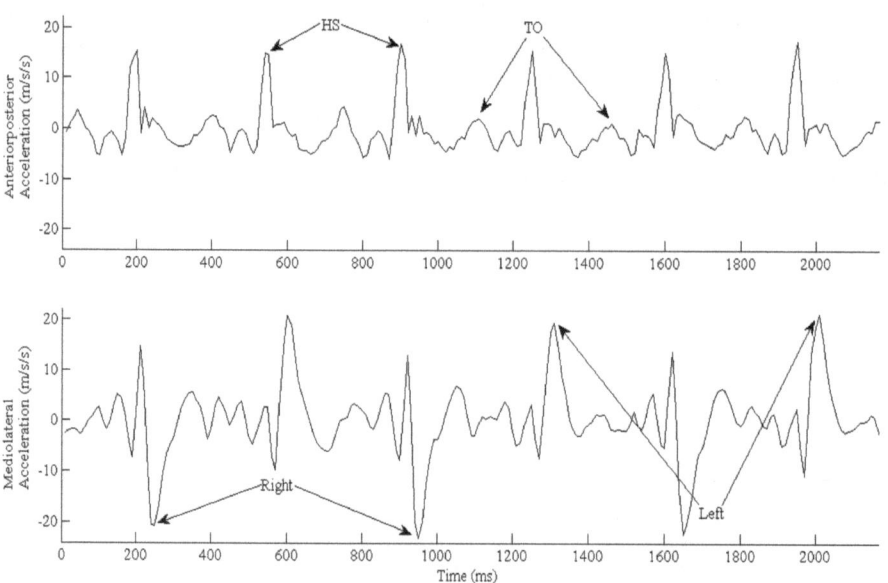

Fig. 4.4 Large peaks, both positive and negative indicate data from the left or right side. Depending on the sensor orientation when placed on the body will dictate whether left or right is in the positive direction. In the case of the data depicted here, it was from the left side of the body

created. This allows progression to be tracked through kinematic changes. If there is an injury through the season, this baseline can be matched, e.g. by overlaying data of different runs to determine differences (Fig. 4.5).

4.3 Swimming

Swimming involves full body coordination. In a technical sense and to put into a comparative perspective, swimming is the most technical of the three disciplines in triathlon events. Up until leading into the 2000 Sydney Olympics, good swimmers tended to dominate competitions, possibly best exemplified by Craig Walton. However, in many competitions, rules were changed in the cycle leg to allow drafting in order for closer finishes of the whole event. This took away a lot of the competitiveness for swimmers and possibly less reliance and time needed by athletes to focus on technique during training sessions. With sensors, armstroke and body roll monitoring is possible. However and especially with arm stroke, identification of event classification within the action is difficult. Fusing sensor and video data makes it possible (James et al. 2012). However, before any typical data collection is possible, a validation analysis is needed.

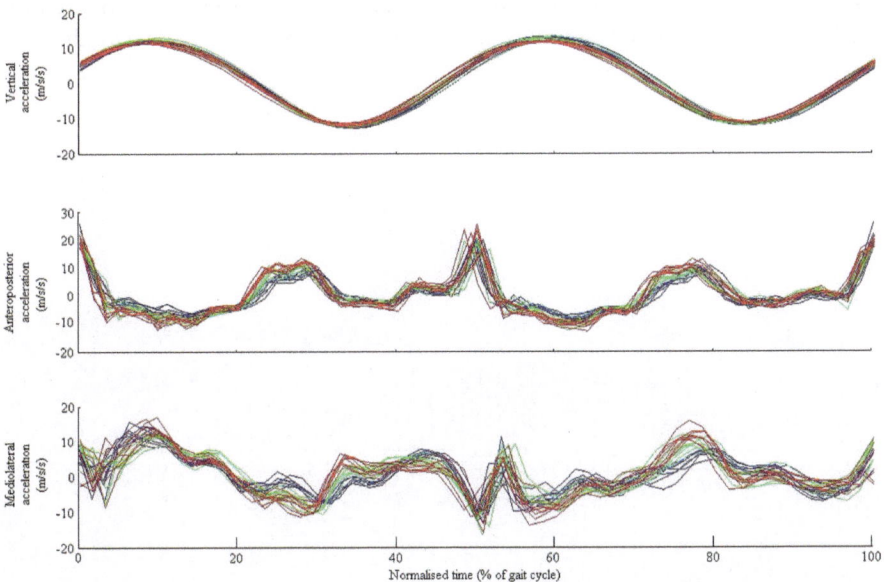

Fig. 4.5 Three separate runs have been taken, and several strides from each run overlayed. The narrower the width of overlay, the more consistent the athlete. The bottom plot is mediolateral and indicates this athlete's sideways movement varies considerably more when compared to their vertical acceleration

To validate inertial sensors in aquatic environments is very difficult, even if the devices themselves are waterproof. The system to compare against, usually the gold standard, is more often than not restricted to a laboratory. For kinematic measures such as arm stroke, infrared 3D motion analysis is a common criterion to compare against. Because of this, early validation required dry land assessment (Fig. 4.6). Another requirement is for supporting technology such as a swimming ergometer to provide adequate resistance to replicate aquatic environments. Additionally, if front crawl (commonly known as freestyle) or backstroke is to be measured, a dry land swim bench needs to be able to accommodate body roll simulations. Previous research has validated sensors in such a way for armstroke classification (Lee et al. 2011).

Because technique has considerable effect in swimming, athlete recruitment typically requires proficient swimmers. This is to ensure that high-quality data profiles to be collected. Generally, competitive swimmers' technique is consistent, and while it may not be symmetrically perfect, their style is in almost all cases consistently the same. In a validation, this can allow for relatively small sample sizes to be effective. For a proof of concept, a single athlete may be all that is needed and at least 10 swimmers should be the target number for a validation. It is important to remember that participants are not the sample size. In a typical sensor validation between two measurement systems, the sample points give the sample size. Therefore, at 100 Hz sampling collection rate, the data quickly accumulates. In this context, what is being tested is whether there is agreement between the two systems (e.g. 3D motion capture and triaxial accelerometer from an inertial sensor). For kinematic event detection, how well do the two systems agree where the same

Fig. 4.6 Dry land validation of inertial sensors measure of arm stroke. Reflective markers have been placed at various points on the participant's arms and torso. One of the 12 infrared cameras used can be seen behind the participant's head

sampling point is. Of numbers from a movement, this can be both magnitude and timing of the data.

For coaches, the typical starting point of an arm stroke is hand entry to the water. From here to hand exit from the water has a general classification of the propulsive phase. Any movement back to the hand entry position classified as the recovery phase. Within the propulsive phase, the hand orientates in slight lateral and medial sweeping action forming an "S" pattern during the pull through the water (Hay et al. 1993). Hand entry is also the initiation of the glide of the hand prior to the downward push of the hand which is the commencement of the propulsive phase of a freestyle arm stroke (Lee et al. 2008). The bulk of arm movement is at the shoulder, and while there are some lateral deviations, the majority of movement is in the sagittal plane, with the propulsive phase predominately arm extension. Therefore, the sagittal plane rotation is around the mediolateral axis of the shoulder joint and results in outputs that can be easily detected in the gyro data that is orientated to this axis.

To provide a water based example of carrying out data collection, here is a case study of how data was collected. It is based on a project looking at different swimming conditions to determine whether different environments impact on sensor outputs. The aim was to determine whether triathlete coaches needed to be specific as to where their athletes should be training. Importantly, this covers areas that should be considered when working with people freely volunteering their time —in other words, "bedside manners".

Triathlon is a sport that was once classified as being made up of three disciplines; swimming, cycling and running. More recently, a fourth has been included, namely the transition. This occurs between each of the others where valuable time can be lost. Depending on the venue, the swim leg can be in a swimming pool, still open water or surf open water.

Most pools are typically designed to accommodate for swimming with lane ropes and lines painted on the pool floor to assist in orientation. Generally, lane ropes are designed to minimise wash from other swimmers as well as chop created by wind affecting the water surface. In this environment, athletes can largely work on their technique for efficient swimming.

Still open water takes the controlled pool type environment away. Therefore, a big issue for an athlete is knowing where to go. Still open water can be environments such as lake, dams, canals and slow flowing rivers. They can be in tidal waters with event timed to coincide with the local tide turning in order to minimise the effect of current. Swimming in a straight line in still open water can be challenging and requires a modification to typical pool freestyle swimming technique. One major change is to swim several strokes before raising the head out of the water to look forward and ensure the correct direction is maintained. This can have an effect on swimming mechanics.

Surf open water largely speaks for itself where the swimming is typically at beaches with the added variable of surf being an issue to contend with. This may be close to shore where waves can gently break through to waves acting more violently, peaking high and "dumping" water in a chaotic manner. Out beyond this hectic zone, the waves may be rolling in with a relatively smooth surface. Or can be affected by wind, causing chop on the surface. Lifting the head for orientation may have to occur more often in surf conditions as opposed to still open water, due to the likelihood of being tossed around in the water.

Therefore, a study was designed to compare the three different aquatic environments that triathletes can encounter. The aim was to determine whether wearable devices could detect changes in arm stroke, body roll and kick kinematics. It was decided that a pilot study was required in order to determine the most effective and efficient way to collect data that had the least impact on anyone volunteering to participate.

The first step was to validate sensor data against an accepted measure. This type of validation needs to be against another motion analysis tool, often 3D motion analysis (Lee et al. 2011). Therefore, the validation was with a swimming ergometer incorporated with a swim bench. This dry land component allowed for direct comparison and enabled a first time preview of armstroke and body roll profiles (Fig. 4.6). Results showed strong agreement between the two systems. This enabled the second phase to be commenced.

The second phase of the study was to recruit one swimmer to carry out the first in-water pilot project. This was in an Olympic swimming pool environment. The purpose of a single swimmer trial is multifaceted. Primarily, there was a need to show that what was found with the dry land study was reflected in the kinematics of actual swimming. There would be little use progressing if the data was not similar and a redesign of the first phase to validate true movement pattern data capture would have been required. Other reasons were to ensure the devices were able to be safely waterproofed and that devices be securely fastened in position. This is for both accurate data capture and not to lose a device—one fallen off in the pool is inconvenient. One falling off in open water may be permanently lost. This also confirmed the chosen sensor positions were effective and meaningful.

The position of the sensors was for one on the dorsal side of each forearm, close to the wrist. This was for armstroke profile and to measure for left and right symmetry in the swimming style. Sensors were also positions at the occipital lobe of the skull (back of the head), the C7 region of the vertebrae (upper spine), at L5/S1 of the vertebrae (lower back) and posteriorly (back) on the mid calf (Fig. 4.7).

The reason for the head and vertebral positions was to measure longitudinal axis (around the spine) torsional rotation as well as any differences in breathing and non-breathing strokes. This was very important due to different breathing patterns of still water (pool) and open water swimming. The head position was difficult to attach due to hair. This was resolved by the swimmer wearing two swimming caps with the sensor secured between the two. On other body parts, sensors were fixed with double sided tape to the skin and for extra security, covered with waterproof tape. Because of the positioning, the lower back sensor had additional security on female participants who wore one-piece bathing suits and depending on the shoulder strap design, sometimes the sensor at the C7 position.

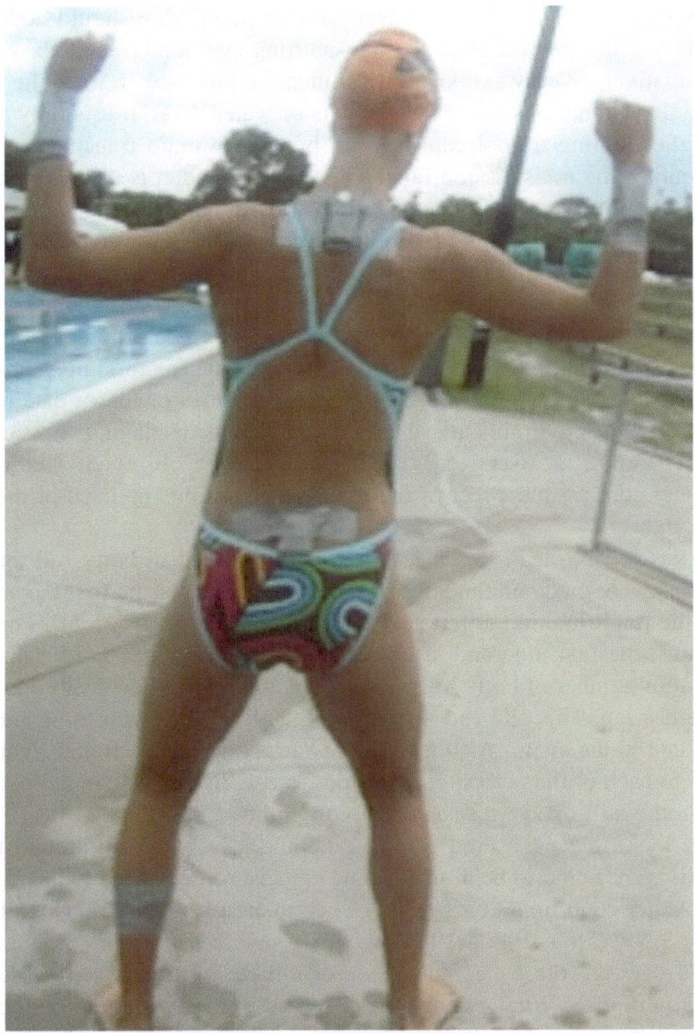

Fig. 4.7 Sensor positions on a swimmer. The sensor positioned at the occipital lobe is hidden by the second swim cap

It is worth highlighting here that a very high level of communication is critical before and during all parts of collecting data. Not only permission to place the sensors on body segments, clear instructions of what is occurring are very important. At this point, volunteers are quite vulnerable due to being in close proximity with someone (the researcher) who they are likely to not know all that well while wearing minimal clothing; therefore, respect is paramount. Continual communication makes sure the athlete is comfortable with you in their "personal space".

Even grabbing an arm to place a sensor without permission should not be carried out, let alone other parts of their body or shifting swimming apparel.

This relatively "easy" pool environment allows to refine the research. Understanding limitations such as the effects of water. Even if it is only a thin film of water after exiting a pool can have an impact on radio transmission capabilities by limiting the transmission range. The need to video is also a consideration. This is to confirm what may occur during a swim. If a camera is used, privacy issues of others may need to be taken into account. In open water, often variations such as stopping or changes in direction can be noted by camera use.

Due to the variations of breathing and raising the head and maybe trunk results in slight changes in data. To best understand what may be differences in the three techniques, one way to easily visually observe the data is to simply print data in kinematic profiles and pin to a wall, by creating columns of sensor data for each of the three different environment's data with each row displaying the same sensor position. Therefore an oversized visual matrix can be observed. This enables to compare and contrast the effect of the different conditions on each of the sensor's kinematic data.

What was looked at in this research was differences in body roll angles. The comparisons were made on the left and right sides, also for the first lap and last lap swum in the pool (looking at left side vs left side and right vs right). For the open water swims, the first nine and last nine body rolls were measured. This reflected the maximum number of body rolls in the pool condition, specifically the right side roll in the last lap. This allowed for body roll early in the swim to be compared to body roll late in the swim. Also from these variations, each left and right side was compared to each of the corresponding sides, e.g. early left side in the pool to early left side in the still open water and surf conditions.

Just by looking at measured values, differences in body roll around the longitudinal axis were found to be different. The greatest roll occurred to the right side in still open water conditions (67°) (Fig. 4.9) compared to surf (61°) (Fig. 4.8) or the swimming pool (58°) (Fig. 4.10). The most visible events that can be seen in the plotted data are the clear indications of tumble turns that can be seen in the swimming pool data. What can also be observed in the swimming pool data is the relative consistency of the left side body roll compared to the right. This indicates an asymmetry in the swimmer's body roll that changes over the duration of the swim (Table 4.1). There was an increase in body roll on both sides with an average of 35° for lap one going to 40° for the final lap and a body roll change of 50–57° on the right for the corresponding laps. Additionally, this means a 15° difference between the left and right sides on the first lap, increasing to 17° on the final lap.

These observances are just part of the story. When statistical analysis are applied (alpha level was set at $p = 0.05$), other inferences can be made (Table 4.2). Statistically speaking, there are differences between the left and right side in all three water conditions, both early and late in the swims. This backs up the reported body roll angles in the previous paragraph. On every occasion, there was a significant difference between the left and right sides. Other variances appear to be less conclusive and consistent. For example, early left and late right side comparisons

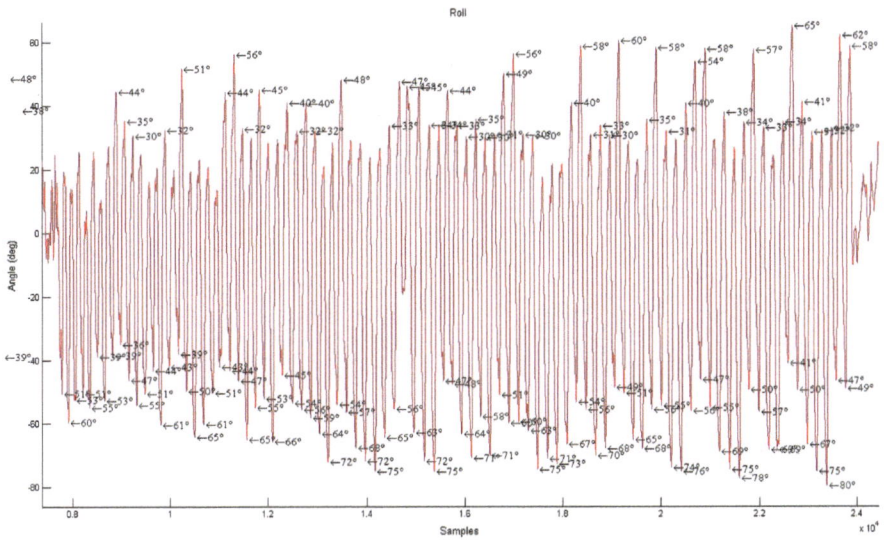

Fig. 4.8 Open surf data. Conditions were moderate with approximately 0.75–1.00 m swell

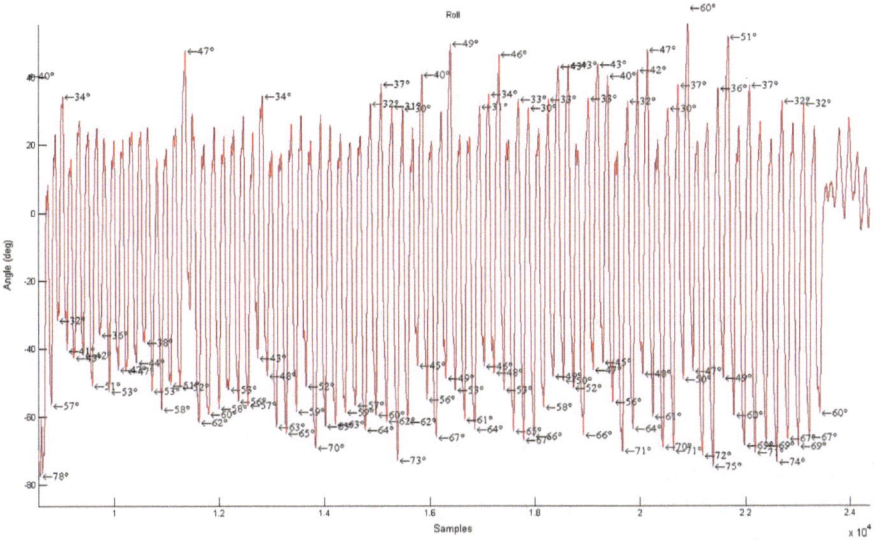

Fig. 4.9 Still open water. Conditions were calm with a slight headwind on the return leg

between the pool and still open water conditions were different. However, early right and late left, there was no significant difference. Further research may be required to determine the reason why, e.g., swimming into surf or with the surf may alter the body roll.

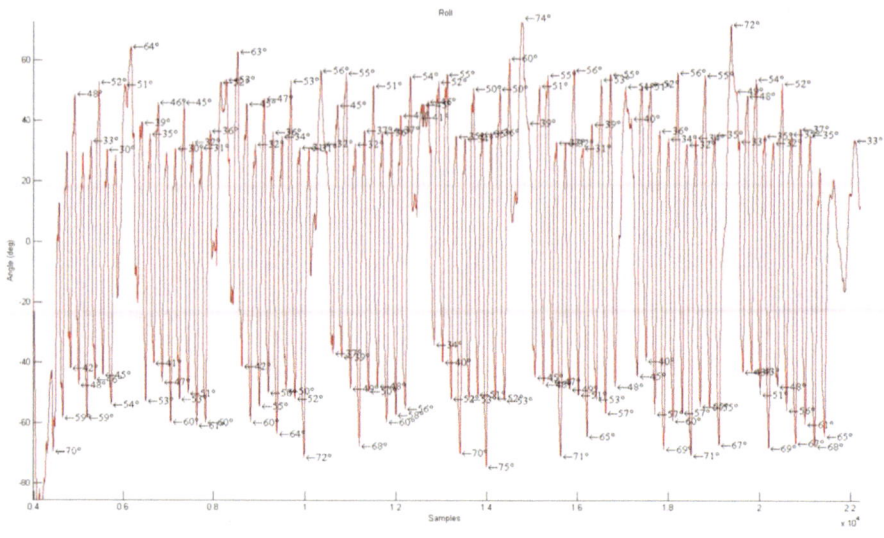

Fig. 4.10 Swimming pool. The pool was situated outdoors

Table 4.1 Magnitude of average body roll and standard deviation for rolling to the left and right side

	Pool		Still open water		Surf	
	Left	Right	Left	Right	Left	Right
Early body roll (degrees)	35.9 ± 9.8	50.4 ± 6.9	25.6 ± 5.5	43.1 ± 8.1	24.0 ± 10.4	48.6 ± 8.4
Difference	14.6		17.6		24.6	
Late body roll (degrees)	40.6 ± 9.2	57.8 ± 10.7	31.8 ± 8.2	67.3 ± 4.7	39.0 ± 14.8	60.8 ± 14.1
Difference	17.2		35.6		21.8	

Early body roll represents measures carried out early in each swim. Late represents measures of body roll late in each of the swims

The other point to comment on is that the swimmer took seven strokes on each side to complete the first lap and eight strokes on the left and nine strokes on the right to complete the final lap, indicating a longer distance per stroke early and a likely influence of fatigue resulting in shorter strokes in the final lap. Shorter stokes are common in later stages of swimming due to the effects of fatigue.

Comparing the open water swims to the swimming pool shows interesting results. Statistically speaking, there are differences between the pool swimming and both open water conditions when only looking at left side body roll.

Table 4.2 T-test results of body roll differences

			Pool				Still Open Water				Surf			
			Early		Late		Early		Late		Early		Late	
			Left	Right	Left	Right	Left	Right	Left	Right	Left	Right	Left	Right
Pool	Early	Left		***0.043**	0.352		**0.035**				**0.036**			
		Right				0.119		0.148				0.632			
	Late	Left				***0.003**			0.056				0.787		
		Right								**0.034**				0.670	
Still Open Water	Early	Left						**0.000**	0.141		0.690			
		Right								**0.000**		0.460			
	Late	Left								**0.000**			0.276	
		Right													0.215
Surf	Early	Left										**0.002**	**0.043**		
		Right													**0.032**
	Late	Left													**0.030**
		Right													

Code:

Left to Right;

Same Side, same Swim, Early to Late;

Same Side, Pool to Still Open Water;

Same Side, Pool to Surf.

Bolded indicates a significant difference. Colour indicates the condition comparison where a significant difference exists

References

J.M. Bland, D.G. Altman, Statistical methods for assessing agreement between two methods of clinical measurement. Lancet, 307–310 (1986)

R. Blickhan, The spring-mass model for running and hopping. J. Biomech. **22**(11–12), 1217–1227 (1989)

N. Davey, A. Wixted, Y. Ohgi, D.A. James, in *The Impact of Technology on Sport II*. A low cost self contained platform for human motion analysis, vol 2, pp 101–111

J.G. Hay, Q. Liu, J.G. Andrews, Body roll and hand path in freestyle swimming: a computer simulations study. J. Appl. Biomech. **9**, 227–237 (1993)

W.G. Hopkins, Spreadsheets for analysis of validity and reliability. Sportscience **19**, 36–42 (2015). www.sportsci.org/2015/ValidRely.htm

D.A. James, R. Leadbetter, N. MadhusudanRao, B. Burkett, D. Thiel, J. Lee, An integrated swimming monitoring system for the biomechanical analysis of swimming strokes. Sports Technol. **4**(3–4), 141–150 (2012). https://doi.org/10.1080/19346182.2012.725410

J.B. Lee, R.B. Mellifont, J. Winstanley, B. Burkett, Body roll in simulated freestyle swimming. Int. J. Sports Med. **29**(7), 569–573 (2008). https://doi.org/10.1055/s-2007-989285

J.B. Lee, R.B. Mellifont, B.J. Burkett, The use of a single inertial sensor to identify stride, step, and stance durations of running gait. J. Sci. Med. Sport **13**(2), 270–273 (2010a)

J.B. Lee, K. Sutter, C. Askew, B. Burkett, Identifying symmetry in running gait, using a single inertial sensor. J. Sci. Med. Sport **13**(5), 559–563 (2010b). https://doi.org/10.1016/j.jsams.2009.08.004

J.B. Lee, B. Burkett, D.V. Thiel, D.A. James, Inertial sensor, 3D and 2D assessment of stroke phases in freestyle swimming. Proc. Eng. **13**, 148–153 (2011)

T.F. Novacheck, The biomechanics of running. Gait Posture **7**(1), 77–95 (1998). https://doi.org/10.1016/S0966-6362(97)00038-6

D.A. Winter, Human balance and posture control during standing and walking. Gait Posture **3**(4), 193–214 (1995)

Chapter 5
Take-Home Messages

5.1 Importance of Meaningful Data

Without meaningful data, assumptions can only be made. Worse still inaccurate information may be provided to athletes and coaches. This can be detrimental to athlete performance through injury or less than ideal training designs. Meaningful data also means that the site chosen for data capture has to reflect what is intended to be investigated, e.g. it is little value looking at sensor data positioned at L5/S1 of the spine if number of kicks were to be measured because kick beat does not match body roll. However, body roll does match arm stroke. Therefore, it could be meaningful to measure armstroke counts from a sensor positioned at the spine. If other arm kinematics were the focus, a spinal mounted sensor will likely be not useful at all and an arm-mounted sensor or sensors is a likely better option. Because of asymmetries in the human body, it is important to determine the depth of understanding that is needed. For example, if only general armstroke information is required, a sensor on one arm may be sufficient. If armstroke kinematics were to be related to body roll, more sensors would be required. This would mean synchronisation of the sensors has to be made. Otherwise, it is difficult to ensure any data collected can be used in any meaningful way. The take-home message here is to make careful decisions of sensor placement, number of devices to use, how to synchronise multiple sensors, what capture frequency to use, and length of data capture. These are just some of the considerations to be made for optimal and effective outcomes for performance or activity monitoring.

5.2 Importance of Accurate Interpretation

Performing data analysis provides a means to confidently state the probability or relationship of an effect on a population from taking a relatively small sample size. However, be for any accurate interpretation can be made, the tools used to collect

© The Author(s), under exclusive license to Springer Nature Singapore Pte Ltd. 2019
J. Lee et al., *Wearable Sensors in Sport*, SpringerBriefs in Applied Sciences and
Technology, https://doi.org/10.1007/978-981-13-3777-2_5

data need to be shown to be valid means and the user has the confidence of its reliability. Like just mentioned in Sect. 5.1, incorrect interpretation can also be detrimental to the end-user. This means that it is important to understand what any data is indicating. Is there a trend? If so and collected from a population sample, how accurate will it be to the whole population? One point that is very important. While a statistical analysis may indicate a correlation, it is not at all effective for identifying the cause (cause and effect) e.g. "The sun rises and morning birds start singing" is a cause and effect. While the same analysis to test whether "The sun rises because morning birds start singing" will give the same correlation. Therefore, it is important to have a sound understanding on the topic or variables being tested in order for sound judgement to be made from the statistical analysis.

5.3   Future Applications

If you are reading this book, chances are you are already using or perhaps wanting to use inertial sensors in human motion studies. Either way, one of the first stages is a kind of discovery stage of the utility of the sensors. Whether you are from primarily a technology-based background "pushing the technology" into an intended application or someone from within the sport sciences "demand pulling" the technology to help solve a problem, you have a complimentary expertise set that you will gradually acquire overtime through your own development of partnering with those from the other perspective. Ideas and approaches can come from either the technology push or the demand pull, and there is no right or wrong way to be doing it.

Over time a concept emerges from that combination of the technology and the application that is often driven by trying to solve a problem, and it is then that turning the concept into something more may start to crystalise. A model to consider following is:

$$Ideation \rightarrow Concept \rightarrow Product \rightarrow Commercialisation$$

Product development is the process that involves taking your idea and turning into a product for wider use. It consists of evaluating it to make sure that it has enough wider appeal to solve other people's problems too, that the problem is sufficiently wide spread enough and that there is enough of a market to develop such a product and a healthy dose of marketing to get there (Crawford 2008). The development of the concept itself has to consider, reliability, that you have enough expertise to take a one off and get it manufacturing ready. One of the greatest challenges at this stage is making sure that the passion for the idea can translate through the rigorous processes of refining and engaging in marketing aspects which are sufficiently exciting as well.

Taking an idea through the process of commercialisation can be confronting as you step back from what you have developed and decide if it is viable, who owns it and how to take it forward. It is here that many good ideas can struggle. Success has many fathers, and you may have involved a lot of people in the development of your idea, deciding how that translates to ownership is often complex especially when institutions are involved. One key stumbling block is equating the future value of an idea or concept with its present-day value. Invention and ideation is a key ingredient in commercialisation, but not the only one as it takes a lot of resources and inputs from many hands to have it transition successfully. One popular approach is to consider setting up a start-up, or start-up like entity (Ries 2011) where a major emphasis is making sure there is heavy interaction with the users or customers, in what as been labelled the product-market fit (Blank & Dorf 2012).

5.4   Conclusions

The aim of the creation of this book is to give a brief overview of the emerging area of wearable technology to assess human movement. It is by no means comprehensive and not intended to be. However, it has provided basic information on how wearables work with a glimpse of the complexities of the technology. Two examples of case studies were given as examples of data collection in two separate environments. Similarities and differences should be seen between the two. The authors aimed to provide basic examples to demonstrate the versatility of technology, along with creating an awareness that depending on what is needed to be measured, specific understandings of research design should be taken into consideration. An introduction to validation statistics along with common, basic test statistics was also provided. These are only a small set of examples of the vast array of statistics designed for data analysis. However, a starting point for any potential user to appreciate those assumptions cannot be made without any data being tested for inferences to be made. Briefly, the authors have finished with some future projections of where wearable technology can be taken and hopefully planting seeds for innovative applications to be designed.

References

S. Blank, B. Dorf, *The Startup Owner's Manual: The Step-by-step Guide for Building a Great Company* (BookBaby 2012)
C.M. Crawford, *New Products Management* (Tata McGraw-Hill Education, 2008)
E. Ries, *The Lean Startup: How today's Entrepreneurs use Continuous Innovation to Create Radically Successful Businesses* (Crown Books 2011)